BioGraffiti
A Natural Selection

BioGraffiti

❁

A Natural Selection

❁

JOHN M. BURNS

W · W · NORTON & COMPANY

New York · London

First published as a Norton paperback 1981

Designed by Tere LoPrete

The illustration of *Drosophila melanogaster* is from
A. H. Sturtevant and G. W. Beadle, *An Introduction to Genetics*
originally published in 1939 and used here with permission by
Dover Publications, New York, from the 1962 edition.

Library of Congress Cataloging in Publication Data
Burns, John McLauren.
BioGraffiti : a natural selection.
Includes index.
I. Title
PS3552.U73246B5   1981   811'.54   80–20875
ISBN 0–393–00031–1 (pbk.)

W W Norton & Company, Inc., 500 Fifth Avenue, New York, N.Y. 10110
W W Norton & Company Ltd., 25 New Street Square, London EC4A 3NT

1 2 3 4 5 6 7 8 9 0

For Sarah

Acknowledgments

Despite revision, certain poems that I wrote to introduce speakers in a Natural History Seminar at the Museum of Comparative Zoology (Harvard University) still relate somehow to an individual, his work, or his talk; and so I want to mention Philip J. Darlington, Jr., Bryan Patterson, Stephen Jay Gould, Robert L. Trivers, Rolla M. Tryon, Jr., Ronald Munson, Joel E. Cohen, Thomas W. Schoener, Michael P. Kambysellis, Theodore W. Pietsch, P. Barry Tomlinson, Thomas Eisner, Douglas E. Gill, and Jane Lubchenco Menge. One short poem complements an n-dimensional lecture that Richard C. Lewontin gave in a team course on evolution. Another borrows from John Ray.

My wife, Sarah N. Burns, did the lioness' share of the slow, difficult job of getting illustrations appropriate to the poems. Ruth E. Hill, Hermine Brand, Esther Reynolds, Leslie A. Garay, Kenneth J. Boss, and Alice F. Tryon helped us in this effort. Three times, when nothing could be found in old books, Robin S. Lefberg gleefully drew the figure desired. A. H. Coleman photographed—and Paula M. Chandoha printed—each illustration with rare craft and patience.

Thank you all.

J. M. B.

Contents

Introduction

Legend has it (it may even be true) that J. B. S. Haldane, when asked by a clergyman what he could infer about God from the works of creation, responded, "He must have had an inordinate fondness for beetles." Were I asked to infer something essential about *Homo sapiens* from his work, I should probably reply that this zoological odd-ball required humor to lighten a life taken too seriously. How else can we explain the fact that every profession has its underground classic of humorous self-deprecation and verse? Garstang's *Larval Forms* has long filled this role for evolutionary biology. But, as a residual Victorian, Garstang turned out some mighty stuffy poems—and recapitulatory theory or the details of invertebrate morphology do not reside on the frontier of modern biology. But voyeurs and hedonists can now rejoice, for John Burns has produced a worthy successor, a work full of all that is modern in evolutionary biology—mathematical modelling, ecological strategies, ethological theories and, oh yes, plenty of sex.

Many of these poems were written for an incongruous setting—the oh-so proper Wednesday luncheon and Natural History Seminar of Harvard University—a convocation called to order by the conventional tinkling of spoon against coffee cup. At least until Burns arrived and started reading his doggerel to lighten the transition from dessert to dissertation.

As we attended the Natural History Seminar ever more faithfully—to hear poems as much as speakers—we came to know the characteristics of Burns' verse—conciseness, outrageous rhyme, more outrageous pun, and a density of allusion that initiated an incongruous style of art—the poem with footnotes (here glossarized).

Who else could say so much with a poem bearing as its entire text the Socratic injunction: "Know/Thyself"? And I have read long, dull chapters on mammalian evolution that contain no more wisdom than Burns' six lines on the "Evolution of Auditory Ossicles": "With malleus/ Aforethought / Mammals /Got an earful /Of their ancestors'/Jaw."

W. S. Gilbert once had the audacity to rhyme "caravanserai" with "wards in chancery," but no one besides Burns would try "aardvark" with "hard bark."

Puns had their golden day. Shakespeare's audiences must have laughed at the first lines of *Julius Caesar* with their plays on "soul" and "sole," "all" and "awl"; now, we are generally embarrassed. Harvard students hiss so automatically at puns that I am almost led to regard the response as the finest example of an innate releasing mechanism in *Homo sapiens*. And Noah Webster did cast punning as "a low species of wit." Yet the suppressed underground of inveterate punners must be far larger than most overtly oppressed minorities. We all have an outcast uncle who puns as others neglect baths or molest babies. But punning is proud (how I wish I could do it)— "It is," as Charles Lamb said, "as perfect as a sonnet." Most of Burns' offerings are pun pure and simple. We laugh or we wince, but the reason is the same—we see ourselves only too well. Addison wrote that "the seeds of punning are in the minds of all men."

Men, it is said, resemble their dogs. Burns is a lepidopterist by trade. Isn't it only just that a student of butterflies should be able to transform the grubby and commonplace into gaudy brightness?

Stephen Jay Gould

MAMMALIA

Conform

I want to be scene,
Not herd, said a wayward
Young springbok, as it split.
Know your place, said
The leader, which is together,
And clubbed the errant back,
Giving it the Gesellschaft.

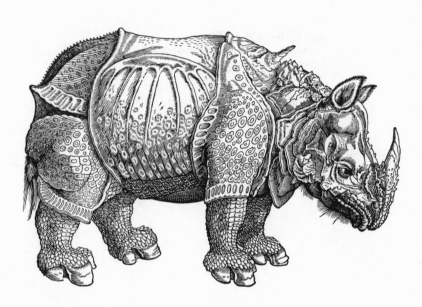

Rhinoceros

Hearing and pre-eminently smell
Make far better sense
To rhinoceros, which sees dimly
(And wears a nosehorn well).
This all but hairless hulk
So enarmored of thick skin with folds
As to lack nonhuman predators of consequence
Thunders toward extinction
Blindly bold to man in self-defense.
Preferentially it holds
Itself apart, hoofing through reeds
And high grass, browsing by dusk
And dawn, solitary in its
Territory save when it breeds.
Communication faintly whiffs absurd:
Movement is action
Movements speak louder than words
Territory marks are piled turds.

Human Evolution in a Nutshell

Once upon
The evolutionary stage
Of anthropoids
Dryopithecus
Was all the rage.

Although some pithy lines
Were later banned
The one that made
The *Gorilla* my dreams
Was merely *Pan*ned.

Meanwhile selected hominids
Foresaw the silver screen.
Australopithecus came close
But *Homo* stole the Holocene.

Beaver
Damn

Let us
Dam
The stupid
Eager
Beaver who
Knowing not
His aspen
From a
Pole in the
Ground
Persists in
Barking
Up
The wrong
Tree.

Oso Negro / Not So Black

The American black bear
Or oso negro
(Now *Euarctos*
Formerly an *Ursus*)
Lumbers through woods
Shredding logs allegro
And slumbers through cold
Dreaming of itself versus
Hordes of wild bees
Their fattening honey
Their stings of outrage . . .

Some western "black bears"
Are cinnamon or gray-blue or white
Which suggests a funny
Coat of many colors
That the species population shares.

Doggerel is Aard Vark

A remarkable beast is the aardvark.
It speaks softly and not with a hard bark.
Some think its song sweeter
But no it ant-eater
The sticky tongue makes it a marred lark.

A Plague o' Both Your Porcupines

Dartmoor, 1969—Two
Porcupines escaped from a private zoo.
Dressed fit to quill they made a prick-
Ly pair, the ♂ up to *Hystrix* . . .
Exponentially their numbers grew.

Devon, five years after—Trees and crops
Are being hit in a nightmare of copse
And robbers by an hysterical rodent plague
Costing residents thousands of dollars. Vague
Grumblings reach the Queen about applying stops.

The Lonely Bull

Pedigreed,
Royally fed and stalled,
Cowed only by their succubus,
I have come to naught
But semen (first class).
Better a detested steer:
Why don't they just
Snatch both
Instead of extracting them
Mechanically
A little at a time?

Hippopotamus amphibius

Much more an enormous pig than a sort of horse,
Hippo lives, as a matter of course,
Both in water—still or running, fresh or salt—
And on adjacent land, where its Gestalt

Takes fifty pounds (dry weight) of grass per night.
In human cropland, which it freely samples,
Much of what it doesn't eat it tramples,
And signs point to a final interspecies fight.

The losing bull in an intraspecific bout
Hides wounded skin and pride in water, where,
With only eyes and nostrils out,
He surveys the scene and takes the air.

To save its skin from air as well as flood,
Hippo "sweats" thick, oily "blood."

Evolution of Auditory Ossicles

With malleus
Aforethought
Mammals
Got an earful
Of their ancestors'
Jaw.

Sabu

Sunstruck from the procreative rut
The sacked nuts of a male squirrel
Eerily retreat along the inguinal canal
To bury themselves above the body
Cavity in a regimen of atrophy.

In good time they recrudesce
And soon are dancing on the coelom
In anticipation of the plunge to
Dry sac, squirrelles, and roundelays.

This cycle is a classic *sabu.**

* An acronym (ascribed to
Down under army officers
In North Africa in World
War II) that stands for
Self-adjusting balls-up.

Equivocus

Because it has
The cheek to broadcast
Bands of black and white,
Zebra seems in hindsight
To be comme un âne
Like a wild ass.

The Baculum

An inarticulate lucky stiff between
Paired spongy corpora casanova,
The baculum (or penis bone) of mammals
Lends firm support to a hard job.

Present in all insectivores,
Bats, rodents, carnivores,
And most primates (but not man),
It comes in many shapes.

That of the walrus (winner of a grand prix)
Is very like a warped baseball bat
Some two feet long. As one old walrus put it,
"Speak softly and carry a big stick."

Animal
Love

Hungering
For your
Dear body
Hugged
More lovingly
Than you
Can bear
I'll eat my
Venus, Hon.

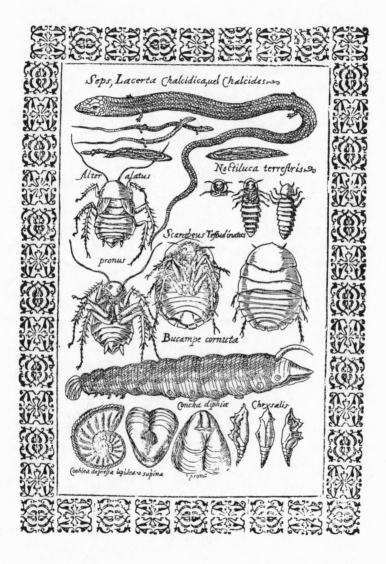

ET ALIA

To a Lonely Hermaphrodite

Know
Thyself.

Lines to an Angler Fish

Aloof and quasi-indestructible
For all the world like some religious sages
The angler ineluctable
Rests on its venter
And passeth for a piece of the rock of ages.

Casting for its bread upon the waters
The phony worm at the end of a dorsal ray
It robs the sea of sons and daughters
Converting, with a violent gulp,
Would-be predators to prey.

Flatfish Development

The bottom-dwelling flounder begins—
Like other fish—as a free-swimmer
With dorsum up, venter down
And left and right both lateral,
But starts revolving by degrees
(Circa ninety) on its long axis
Til sides are top and bottom.
Meantime the lower eye moves topside
Over the head where the once dorsal
Fin grows forward, and the down side blind side
Becomes a white side while the eyed side
Colors up with spots that have
The look of eyes all over the plaice.

Competition and Catastrophe in Polliwogs

In the clutch, frogs overload a temporary pond
Egging ever more offspring on to meta-
Morphosis and the best of both worlds.
Developing tads (who cannot know the dad
That amplexed mom, whoever she is)
Graze the indelicate bloom from a green alga
While the perimeter of their first world shrinks.
A mixed bag. The size of individuals
Creeps up, their numbers plunge, and
Proportions of species and frequencies of genes
Shift. Too soon this semi-selected microcosm
Dwindles to a viscous soup of writhing macrosperm—
A mammoth ejaculate going down the wrong tubes.

An Olé for Anolis

The male of an anole named *garmani*
Is subject to sexual selection
As he seeks an existence of harmony
And the chances to make a connection.

He maneuvers for adequate holdings
(O plot for becoming attractive!)
By resorting to dewlap unfoldings,
Being robust, and frightfully active.

He both acts the competitive wizard
And succeeds in enticing a friend
Who will mate him. But, being a lizard,
He does it by halves in the end.

Snake Pit

Even in darkness
With only the data
From an obliquely forward-looking
Temperature-sensitive pit
Twixt nostril and eye
A toxic crotalid
Computes the course
Of a passing mouse
And makes the spectral leap
From infrared to ultraviolent
With double pinpoint precision.

The Hare and the Tortoise

We as much as the insufferable hare
Stared one to another in unuttered
Disbelief at the tortoise's dare.
Recovering his tongue, the hare served up
His scorching brand of mock turtle fare
As we prescribed a racing course
That ended where it began, there
At the trumpet vine.
                              They were off . . .
Our short wait lengthened through glare
To dusk when we strained to see the winner
Coming tortoise. Fair and square
He had plodded through the smug sleep
Of the hairy braggart who, in a scare,
Now tore to second place, shouting
S.O.B. (or some other macronym),
And slunk away muttering Snafu
(An acrimonious anachronistic acronym).

The Great Tit Rip-off

Left on British doorsteps early in the day
Bottles of milk capped with foil or cardboard
Wait to be taken in and put away.

No one thought this quintessentially
Mammalian food was for the birds
But no one reckoned with the local Paridae:

Four species quickly learned to use the bill
To puncture, peel, or highly prize the caps—
Thus ultimately foiled—and swill.

This ethologic shift involving diet
Smacks (in hindesight) of the loosely preadaptive
Inasmuch as tits were the birds to try it.

Niche
Splitting

Nuthatch
Works
A tree
Trunk
Down

Creeper
Does
It
Up
Brown

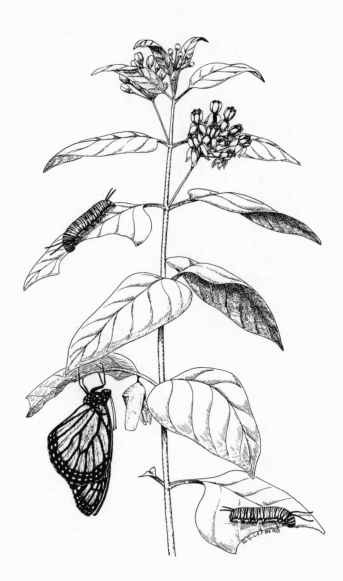

Up the Food Chain

Things aren't always what they aposeme:
The model monarch as a caterpillar eats
Milkweeds, stores their hearty poisons,
    and the butterfly defeats
Some portion of an avian predation team
By advertising—with show of color, lack of haste—
That those that dare to peck will find it in bad taste;
But the monarch straddles defence with another gimmick
When it feeds on milkweed species in which there are
No poisons and becomes an edible adult,
    the so-called "automimic."
(Or is that a chrysomelid beetle that recalls
    the German peoplescar?)

Lethal Yellowing

No oiled and beaten path connects
The harrowed groves of academia
With collapsing rows of macadamia
Or coconuts. Tropical trees toppled
To earth by some intemperate pathogen
Attract few scientific bees
Hell-bent on conquering disease.
Workers are deterred by a monumental
Dearth of basic information
On anatomies and physiologies.
Crops such as palms have not been studied
Up and down like, say, wheat, whose attributes
Glut many a professional serial publication.
Is it any wonder, then, that *Cocos*
Planters of the Caribbean bellow
When they see their acreage succumb
To lethal yellowing, "The grass
Is always greener on the other fellow"?

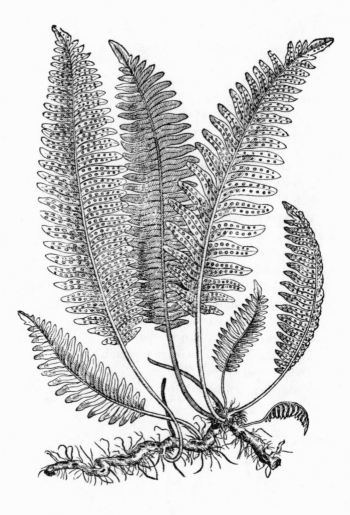

One Good Fern Deserves Another

A tree may be a prologue when it has a hyper bole.

Prothallia of ferns are always haploid
Producing sperms and eggs that seize the procreative role
When, of a dampness, they unite to form a diploid.

Up springs the frondly sporophyte,
      with rhizome, root, and rachis
And a meristem that's apical and tight.
It uncoils; but on a leaf that is preparing for meiosis
Sporangia in clusters make a very sori sight.

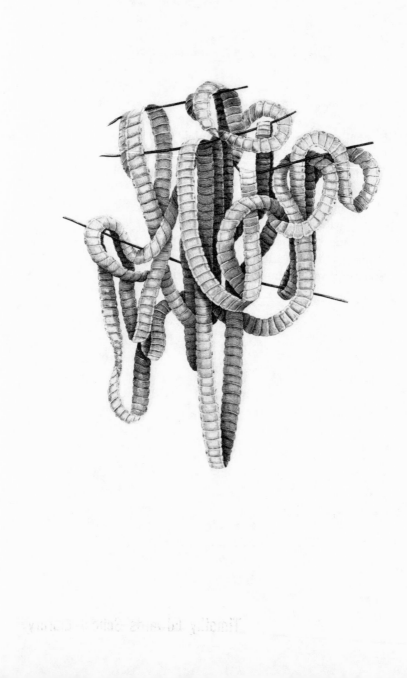

Gutless Wonder

Though lacking skeletal strengths
Which we associate with most
Large forms, tapeworms go to great lengths
To take the measure of a host.

Monotonous body sections
In a limp mass-production line
Have nervous and excretory connections
And the means to sexually combine

And to coddle countless progeny
But no longer have the guts
To digest for themselves or live free
Or know a meal from soup to nuts.

GENERALIA

Lake

Between two states of matter
A line of symmetry:
Hemlocks dark above in spired reality
Are mocked below by colder forms
Disposed to flatter.

Three dimensions stripped of one
Become ideal.
Where crowding rhododendrons kneel
Soft white trusses flushed with pink and sun
Reflect on cool perfection.

Corollas tarnish, disengage and fall
Rippling an image
In a crowning show of futile homage.
Turning sodden brown, they sink through surface
Into not at all.

Desert Range

My faults are quiet now.
I rest in sprawling sunlight
And dusty devils try my dirty flanks
While I try to recollect the heightened
Spirits of my youth. How

If those angels saw my angles now
Bahada-bathed in soiled rock
Would they still sing? Could they
Do anything but mock?

For naturally I gave myself away
To fanfare, smiling at first as it caressed
My foot. But then

My fans arose and coalesced
To my own detriment. And still I must endure

This wearing slow return to ancient orogen.

Drosophila in Paradise

Far east of the china plate
The earth breaks, leaks, and clots
And repeats . . .
Fire into water; earth into air.

In the main the islands come and go
Forming an archipelago.

Endless emigrants from living land
Stream forth by sea and air to found
Or founder . . .
A striking and attenuated few
Trickle their bits from a wealthy heritage
Into half-baked gene pools
About this insulated and eclectic range,
A patchwork world that naturally selects
In new directions its kaleidoscope
Of species in genetic revolution.
The hotstage of ultrarapid evolution
Awaits the leading player.

Enter from afar the gravid female fly
Or a pair of dipterans that fly united
Ours is not to reason why

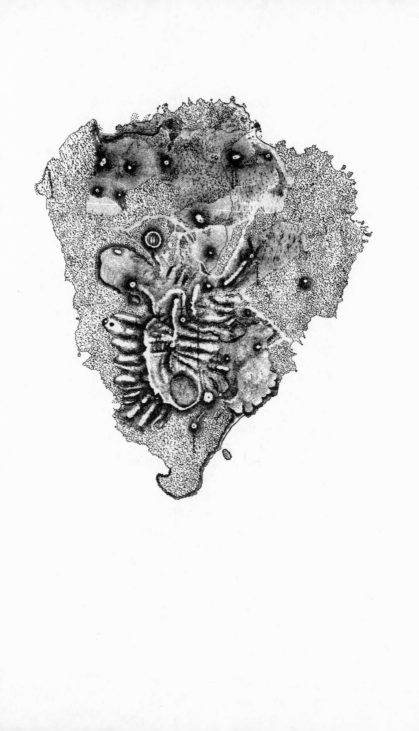

But here they are
With everything they could have wanted:
Beaucoup de Lebensraum
A steamy potpourri of plants.
(Although these flies herbivorous
Would rather phyte than switch,
In a future evolutionary hitch
Some will—to spider eggs!)

Inbred offspring swell the beachhead
Generate their own invading force
And penetrate erratically
A monolithic ecologic vacuum.
To some, put off by numbers,
Falls the vagrant lot;
An offbeat drifter takes another isle . . .
With reproductive ties that bind
Dissolved by water in between,
The replicated colonies
Evolve in part in parallel
But mostly come to see
Sporadic interimmigrants
No longer as their kind.

So species in division multiply.
Among the shifty lava flows, the island flies
Continually colonize and differentiate,
Settling back from time to time to coexist.
Cycle after cycle amplifies
This rich endemic fauna to create
A matchless, if outlandish, marbled layer cake
Mushroomed out of all expected size and shape.
Though countless pieces have gone down
The lubricated gullet of extinction,
There are (by latest estimate)
In the Hawaiian Islands
Some seven hundred
Drosophilid
Species of
Distinc-
Tion.

Fitness By Any Other Name
Would Be As Loose

A group inept
Might better opt
To be adept
And so adopt
Ways more apt
To wit, adapt.

Rocky Intertidal Strife

Fighting the ups and downs
Of an ever-lavin'
Blue-dyed
Wateredge
The rocky intertidal
Lathers its interface
And from time to time
Appears
To hold the sea at bay
With a show of mussels.
This prime frontage
Zoned
For crowded uneasy living
Supports a variably piled
Mat
Of animals and plants
Inclined to be exclusive.
I have heard green algal
Seaweeds such as *Ulva* say
"Like good producers, we eat
Light
But the goddam periwinkles
Littorally graze us—and *Fucus*—
In a catatrophic way."

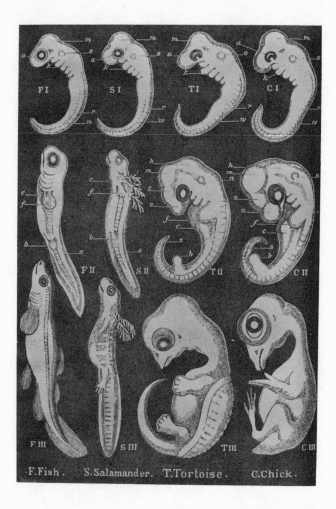

FI · SI · TI · CI

FII · SII · TII · CII

FIII · SIII · TIII · CIII

F.Fish. · S.Salamander. · T.Tortoise. · C.Chick.

Ontogeny and Phylogeny

In the beginning is the end;
But ends unfold, becoming strange.
Lives—and generations—suffer change.
The tested metabolic paths will tend
To last and shape the range
Of future evolution from the past.

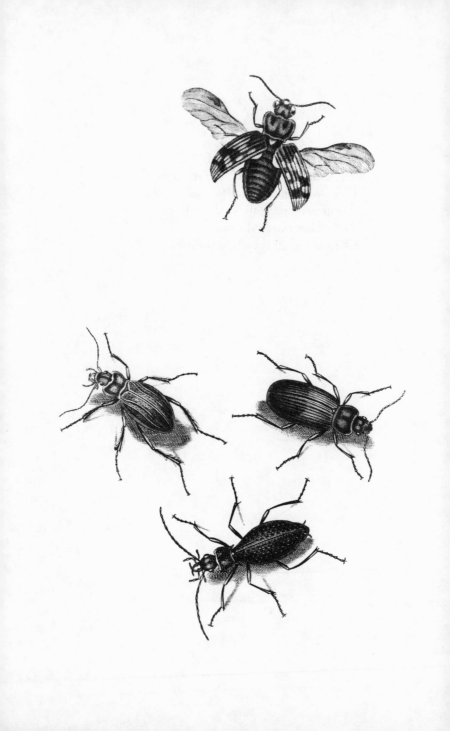

Homage to P. J. Darlington, Jr.

Carabids are beetles of ground.
So spots where carabids are found
Are grounds for inferring
That there they're occurring.
This circular reasoning is round.

Without wings they're more apt to stay there.
But the wingèd may take to the air
Dispersing in myriads
Through Tertiary periods.
We know they all started, but *where*?

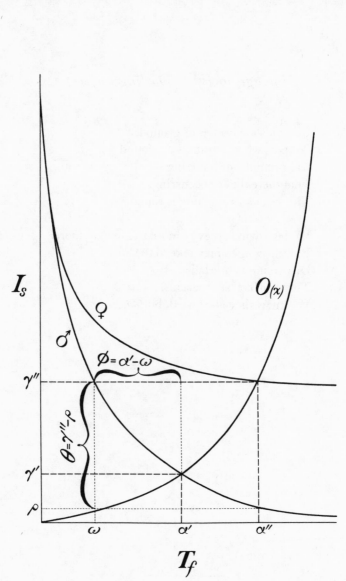

Biomodels

A model in its elegance
Is better than reality
Its graphical simplicity
Denotes a rare intelligence.

The simple graph incites the wrath
Of field men who, half undressed,
Go rushing out to start a test
Which culminates in aftermath.

The Last Rite of
Sylvan Spring

Pink lady's-slippers
And nonrustic beer cans
Garnish the litter of leaves
A mere stone's Thoreau
From Walden Pond.

Province of Biology

Though scientist and artist both create,
The latter makes the better loner.
That each is bound to be a goner
Is not a lively subject for debate.

Simpson said taxonomy is art as much as science.
By most biologists, the systematics is eschewed
And rarely swallowed; for some, a thoughty food
Involves a mix of math and life in an ethereal alliance

About as holy as a sieve. If all of life without and in shall
Not be readily reducible to quantitative theory,
It may be fair that a philosopher should query
"Is the science of biology provincial?"

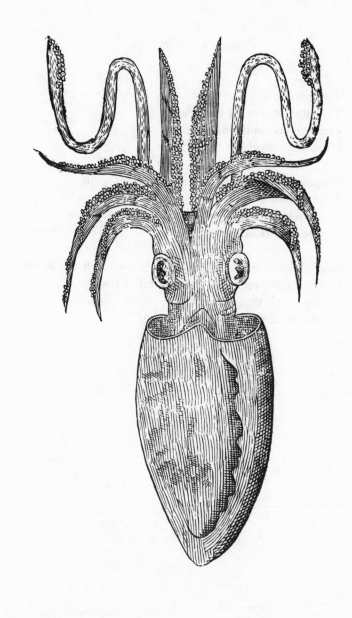

Elucidation
Blues

With a plethora
Of words
The would-be
Explicator
Hides himself
Like a squid
In his own ink.

In Temperate Spring

The air distinctly out of focus:
Everywhere *Forsythia.*

Slopes a wash of *Ceanothus;*
Grassland with *Eschscholzia.*

Cornus and azalea.
Arctostaphylos at Idria.

Like the California oak-moth
I'm a mindless *Phryganidia.*

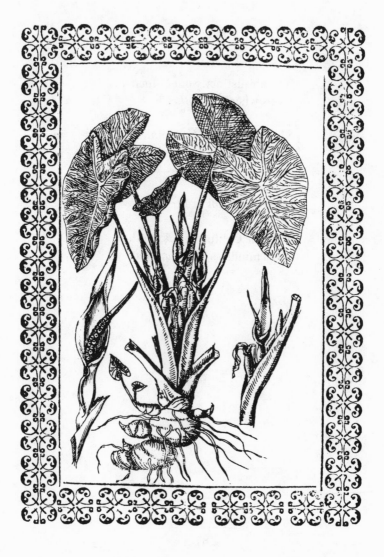

❀

MARGINALIA

❀

Glossary and Notes

Alluvium—Soil, sand, gravel, rock, or the like deposited by running water. At the mouth of a steep, narrow valley or canyon, such stream-borne material spreads out to form an alluvial fan.

Amplex—Verb derived from the noun amplexus: the prolonged mating embrace of a male frog or toad in which he fertilizes the female's eggs as they are extruded.

Âne—French noun meaning donkey, ass, blockhead.

Anolis—A large genus of New World iguanid lizards, commonly called anoles (but also called American chameleons because many species can change color).

Aposeme—Derived from the adjective aposematic, which refers to coloration of a warning (and hence conspicuous) nature evolved by unpalatable, noxious, or poisonous animals.

Arctostaphylos—An almost wholly North American genus of shrubs (family Ericaceae) usually having evergreen leaves and smooth mahogany-colored bark; commonly known as manzanita.

Auditory ossicles—In the middle ear of mammals, three small sound-transmitting bones (malleus, incus, and stapes) evolutionarily derived from skeletal jaw supports (in turn, derived from gill supports).

Australopithecus—A genus comprising one or two (or a very few) extinct species of hominids known from Old World fossils of Pleistocene age. *Australopithecus* is believed to have given rise to *Homo*.

Bahada—A continuous slope of alluvium (q.v.) flanking a mountain range and formed by lateral union of many adjacent alluvial fans.

Bole—The trunk of a tree.

By halves in the end—Male lizards (and snakes) have paired, saclike, eversible copulatory organs—called hemipenes—only one of which is used at a time.

Ceanothus—A North American genus of much-branched, showy shrubs or small trees (family Rhamnaceae) with many far western species, including buckbrush, blue-blossom, and mountain-lilac.

Cocos—That genus of palm trees which includes the coconut palm.

Coelom—The body cavity of many kinds of animals, including vertebrates.

Cornus—A North Temperate genus comprising chiefly shrubs or small trees (family Cornaceae) called dogwoods.

Corpora casanova—Derived from corpora cavernosa: elongate bodies of spongy fibrous tissue (distensible with blood) forming the erector set of a penis.

Crotalid—A snake belonging to the Crotalidae, a family of heavy-bodied poisonous species—rattlesnakes, moccasins (e.g., copperhead, cottonmouth), bushmaster, fer-de-lance, and others—collectively known as pit vipers.

Darlington, P. J., Jr.—Evolutionary biologist, zoogeographer, and beetle taxonomist (partial to Carabidae or ground beetles) at the Museum of Comparative Zoology.

Dewlap—In *Anolis* lizards, a fold of loose skin that can be extended downward from the midline of the throat like a fan; often conspicuously colored and patterned, the dewlap is used in territorial and courtship displays.

Diploid—Having two sets of chromosomes. (This is usual for cells of an individual that develops from a fertilized egg.)

Dipterans—Flies; insects of the order Diptera—so named because they have only two wings (rather than four, which most flying insects have).

Dorsal ray—Any of a number of slender rods extending upward along the midline of the back of a fish where they usually help support a fin.

Dorsum—Back.

Dryopithecus—A genus comprising several extinct species of anthropoid apes known from Old World fossils of Miocene and Pliocene age. *Dryopithecus* is thought to be ancestral to *Pan* and, perhaps, to other anthropoids, as well as to early hominids.

Equus—The genus that includes horses, zebras, asses, and their kin—the surviving members of the horse family.

Eschscholzia—A genus of western North American poppies that includes the California poppy.

Ethologic—Behavioral.

Fucus—A worldwide genus of brown algae inhabiting the mid rocky intertidal zone.

Gamete—An egg or a sperm.

Gesellschaft—German noun meaning association, company, society, fellowship, club.

Gestalt—German noun meaning form, figure, shape, frame, stature.

Haploid—Having a single set of chromosomes.

Hearty poisons—Cardiac glycosides.

Hindesight—See J. Fisher and R. A. Hinde, The opening of milk bottles by birds, *British Birds* 42: 347–357 (1948).

Hippopotamus—"Riverhorse"; derived from two Greek words: hippos (horse) and potamus (river).

Holocene—Or Recent: in the geologic time scale, the latest epoch, extending from the end of the Pleistocene (perhaps 11,000 years ago) to the present.

Hystrix—A genus of Old World porcupines.

Inguinal canal—The open passage that persists between the scrotal sacs and the abdominal cavity in males of certain mammals. Through this canal the testes withdraw to the interior between breeding seasons.

Lethal Yellowing—See P. B. Tomlinson, Lethal yellowing of coconut—the importance of basic research, *Fairchild Tropical Garden Bulletin* 27 (4): 7–12 (1972).

Macronym—A neologism meaning "big name."

Malleus—The outermost of a chain of three little sound-conducting bones in the mammalian middle ear; one end of the malleus is against the eardrum.

Meiosis—Two special, sequential cell divisions that reduce the number of sets of chromosomes in the dividing cells by half—usually from two sets (the diploid condition) to one (the haploid condition).

Meristem—Undifferentiated plant tissue whose cells can continue to divide, some daughter cells differentiating into specialized tissues but others remaining undifferentiated to perpetuate the meristem.

Ontogeny—The developmental history of an individual from beginning (fertilization) to end, with emphasis on earlier stages.

Orogen—A region of mountain-building activity.

Oso negro—Spanish for black bear (*Euarctos americanus*).

Ossicles—See Auditory ossicles.

Pan—The genus that originally contained only chimpanzee—but that now includes both chimpanzee and gorilla. The gorilla used to be in genus *Gorilla* which became a synonym of *Pan* (the older name) when specialists judged that chimpanzee and gorilla are so closely related that they properly belong in the same genus.

Paridae—A family of small, plump, highly active songbirds including chickadees, titmice, and tits.

Phryganidia—The only genus of dioptid moths occurring in the United States: a single species (*Phryganidia californica*) inhabits California, where it feeds as a larva on oaks.

Phylogeny—The evolutionary history (ancestral-descendent relationships) of a group of organisms. (The "group" may vary in size from a few related

---

97

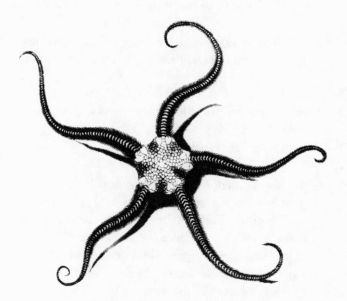

species within a single genus to all of the living things that have ever appeared on earth.)

Phyte—A combining form meaning plant.

Plague o' Both Your Porcupines, A—See British fighting a plague of porcupines in Devon, *The New York Times*, Sunday, October 27, 1974.

Plaice—Any of various European and American flounders.

Prothallia—Plural of prothallium: in the life cycle of ferns, the haploid, gamete-making individual that develops from a spore.

Rachis—The central axis of a fern leaf or frond.

Rhizome—A horizontal, underground stem of a plant.

Simpson, G. G.—Evolutionary biologist, geologist, vertebrate paleontologist, and mammal taxonomist at the American Museum of Natural History and Columbia University, then the Museum of Comparative Zoology, and finally the University of Arizona.

Sori—Plural of sorus: a group of sporangia on a fern leaf.

Sporangia—Plural of sporangium: a case containing spores.

Sporophyte—In the life cycle of ferns, the relatively conspicuous diploid individual that makes spores by meiosis.

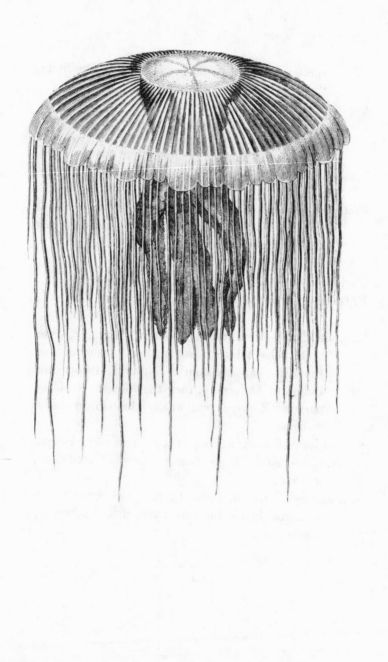

Systematics—Lately redefined as the study of organismic diversity in all its aspects, but, in loose usage, synonymous with taxonomy (q.v.).

Taxonomy—The theory and practice of classifying organisms.

Tertiary—In the geologic time scale, the first (and much the longer) of the two periods of the Cenozoic Era. The Tertiary extends from the beginning of the Paleocene, about 70 million years ago, to the Pliocene/Pleistocene boundary.

Truss—A compact flower cluster at the end of a stem.

Ulva—A worldwide genus of green algae inhabiting intertidal and subtidal zones and commonly known as sea lettuce.

Venter—Belly.

Sources of Illustrations

16	Faber, Johannes. *Animalia mexicana, Descriptionibus, Scholiisq. Exposita.* Romae: Apud Jacobum Mascardum, 1628.
18	As for p. xii.
20	As for p. xvi, except Vol. 8. 1760.
22	As for p. xvi, except Vol. 7. 1758.
24	As for p. 2.
26	As for p. xii.
28	As for p. 12.
30	Colonna, Fabio. *Minus Cognitarum Rariorumque Nostro Coelo Orientium Stirpium . . . .* Roma: Jacobus Mascardus, 1616.
32	Aldrovandi, Ulysse. *Ulyssis Aldrovandi . . . De Reliquis Animalibus Exanguibus Libri Quatuor, Post Mortem eius Editi: Nempè de Mollibus, Crustaceis, Testaceis, et Zoophytis.* Bononiae: Apud Jo. Baptistam Bellagambam, 1606.
34	Bleeker, Pieter. *Atlas Ichthyologique des Indes Orientales Néêrlandaises.* Planches. Vol. 5. Amsterdam, "1865" [1869].
36	As for p. xii.
38	Shaw, George. *General Zoology, or Systematic Natural History.* Vol. 3. Part 1. *Amphibia.* London: G. Kearsley, 1802.
40	Duméril, M. Auguste, and Bocourt, M. *Mission Scientifique au Mexique et dans l'Amerique Centrale.* Vol. 1–5. Paris, 1870.
42	As for p. xii.

44 As for p. 2, except Vol. 3. *Reptiles, Fishes, Molluscs.* London: Routledge, Warne, and Routledge, 1863.

46 Butler, Arthur G. *Birds of Great Britain and Ireland; Order Passeres.* Vol. 1. Hull and London: Brumby & Clarke, [1896–1899].

48 Butler, Arthur G. *British Birds with their Nests and Eggs.* Vol. 1. Hull and London: Brumby & Clarke, [1896].

50 Drawn by Robin S. Lefberg.

52 Treloar, W. P. *The Prince of Palms.* London: Sampson Low, Marston, Searle, & Rivington, 1884.

54 Mattioli, Piétro Andrea. *Commentariorum in VI libros Pedacii Dioscoridis de Medica materia.* Venetiis: Apud Felicem Valgrisium, 1583.

56 As for p. xii.

58 Haeckel, Ernst. *Die Radiolarien (Rhizopoda Radiaria). Atlas.* Berlin: Georg Reimer, 1862.

60 Bunnell, Lafayette H. *Discovery of the Yosemite, and the Indian War of 1851, which Led to that Event.* Chicago: Fleming H. Revell, 1880.

62 Drawn by Robin S. Lefberg.

64 Bishop, Isabella L. (Bird). *The Hawaiian Archipelago. Six Months among the Palm Groves, Coral Reefs and Volcanoes of the*

Sandwich Islands. London: John Murray, 1875.

66 Darwin, Charles. *Geological Observations on the Volcanic Islands, Visited during the Voyage of H.M.S. Beagle . . . .* London: Smith, Elder and Co., 1844.

68 Sturtevant, A. H., and Beadle, G. W. *An Introduction to Genetics.* 1939. Reprint. New York: Dover Publications, 1962.

70 Reade, Winwood. *The African Sketch-Book.* Vol. 1. London: Smith, Elder, & Co., 1873.

72 List, Theodor. *Die Mytiliden.* In *Fauna und Flora des Golfes von Neapel und der angrenzenden Meeres-abschnitte.* Berlin: R. Friedländer & Sohn, 1902.

74 Haeckel, Ernst. *The Evolution of Man: A Popular Exposition of the Principal Points of Human Ontogeny and Phylogeny.* Vol. 1. New York: D. Appleton & Co., 1879.

76 As for p. xii.

78 Drawn by Robin S. Lefberg.

80 Cornut, Jacques. *Canadensium Plantarum, aliarúmque nondum editarum Historia. Cui adjectum est ad calcem Enchiridion Botanicum Parisiense, Continens Indicem Plantarum, quae in Pagis, Silvis, Pratis, & Montosis juxta Parisios locis nascuntur.* Parisiis, 1635.

82 As for p. 58.

84   As for p. 32.

86   Emory, William H. *Notes of a Military Recon-
        noissance, from Fort Leavenworth, in
        Missouri, to San Diego, in California, in-
        cluding parts of the Arkansas, Del Norte,
        and Gila Rivers.* Washington: Wendell and
        Van Benthuysen, Printers, 1848.

88   As for p. 30.

90   As for p. xii.

94   As for p. xii.

98   As for p. xii.

100   As for p. xii.

102   As for p. 30.

108   As for p. xii.

Index of First Lines and Titles

———————————

110